Joseph Abruscato   Joan Wade Fossaceca   Jack Hassard   Donald Peck

HOLT SCIENCE

Holt, Rinehart and Winston, Publishers
New York · Toronto · Mexico City · London · Sydney · Tokyo

THE AUTHORS

Joseph Abruscato
Associate Dean
College of Education and Social Services
University of Vermont
Burlington, Vermont

Joan Wade Fossaceca
Teacher
Pointview Elementary School
Westerville City Schools
Westerville, Ohio

Jack Hassard
Professor
College of Education
Georgia State University
Atlanta, Georgia

Donald Peck
Supervisor of Science
Woodbridge Township School District
Woodbridge, New Jersey

Cover photos, front: M.A. Chappell/Animals Animals; back: George Holton/
Photo Researchers, Inc.
The tufted puffins shown on the front and back covers inhabit islands off the coast
of Alaska. Puffins can also be found in other parts of the north Pacific coast
and the North Atlantic. These birds are expert divers and underwater swimmers.

Photo and art credits on page 231

ACKNOWLEDGMENTS

Teacher Consultants

Armand Alvarez
District Science Curriculum Specialist
San Antonio Independent School District
San Antonio, Texas

Sister de Montfort Babb, I.H.M.
Earth Science Teacher
Maria Regina High School
Uniondale, New York
Instructor
Hofstra University
Hempstead, New York

Ernest Bibby
Science Consultant
Granville County Board of Education
Oxford, North Carolina

Linda C. Cardwell
Teacher
Dickinson Elementary School
Grand Prairie, Texas

Betty Eagle
Teacher
Englewood Cliffs Upper School
Englewood Cliffs, New Jersey

James A. Harris
Principal
Rothschild Elementary School
Rothschild, Wisconsin

Rachel P. Keziah
Instructional Supervisor
New Hanover County Schools
Wilmington, North Carolina

J. Peter O'Neil
Science Teacher
Waunakee Junior High School
Waunakee, Wisconsin

Raymond E. Sanders, Jr.
Assistant Science Supervisor
Calcasieu Parish Schools
Lake Charles, Louisiana

Content Consultants

John B. Jenkins
Professor of Biology
Swarthmore College
Swarthmore, Pennsylvania

Mark M. Payne, O.S.B.
Physics Teacher
St. Benedict's Preparatory School
Newark, New Jersey

Robert W. Ridky, Ph.D.
Professor of Geology
University of Maryland
College Park, Maryland

Safety Consultant

Franklin D. Kizer
Executive Secretary
Council of State Science Supervisors, Inc.
Lancaster, Virginia

Readability Consultant

Jane Kita Cooke
Assistant Professor of Education
College of New Rochelle
New Rochelle, New York

Curriculum Consultant

Lowell J. Bethel
Associate Professor, Science Education
Director, Office of Student Field Experiences
The University of Texas at Austin
Austin, Texas

Special Education Consultant

Joan Baltman
Special Education Program Coordinator
P.S. 188 Elementary School
Bronx, New York

CONTENTS

CHAPTER 1

ANIMALS ARE DIFFERENT

1.

MAMMALS

All of these animals are **mammals.**
How are they the same? All mammals
have **hair** or **fur.** The hair helps them
keep warm. Mammals are **warm-blooded**
animals. The weather may be cold or
warm. But their bodies always stay the
same. Even when it is cold outside, their
bodies are warm.

All mammals are born alive. The babies drink milk from their mother. Mother mammals take care of their babies. All mammals breathe air with their **lungs**.

Mammals are different in some ways. They eat different foods. The mother dog eats meat. Some mammals eat only meat. Lions eat only meat.

Rabbits eat only plants.

Raccoons eat both plants and small animals.

ACTIVITY

What do mammals eat?

1. Find pictures of mammals.

2. Write each animal's name.

3. Put the mammals into 3 groups.

4. Which ones eat animals?
   Which ones eat plants?
   Which ones eat plants and animals?

2.

BIRDS

An eagle is a **bird.** Birds are animals with **feathers** and **wings.** The eagle's feathers help keep it warm and dry. The large wings help it to fly. What other birds do you know?

Birds are the same in many ways. All birds lay eggs. The mother bird lays her eggs in a nest. The mother or father sits on the eggs. This keeps them warm.

Soon babies hatch from the eggs.
The parents take care of their babies.
Do you know how?

All birds are warm-blooded. Even in winter, their bodies are warm. Birds breathe air with their lungs. How are birds and mammals the same? How are they different?

Look at all the birds again. They all have **beaks.** Are all the beaks the same shape? How does each bird use its beak?

ACTIVITY

Look at some feathers.

1. Collect some feathers.

2. How does each feather feel? How does each feather look?

3. Drop some water on each feather.

4. Does the water go through the feathers?

5. How do feathers help a bird?

3.

FISH

Many different kinds of animals live in water. **Fish** are animals that live in water. They use their tails and **fins** to swim. Some fish live in rivers and lakes.

Many fish live in the ocean. The seahorse is an ocean fish. It swims upright in shallow water. Do you see the fins and tail on this fish?

Fish are special water animals. They
are covered by hard **scales.** The scales
help protect the fish.

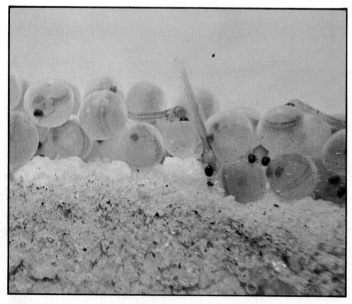

Most fish are born
from eggs. Most fish
do not take care of
their young. How are
fish different from
birds and mammals?

Fish are **cold-blooded**. Their bodies
are not always warm. Fish that live in
warm water have warm bodies. Fish
that live in cold water have cold bodies.

All fish have **gills.**
A fish uses its gills to
breathe. The gills
take air from the
water. The gills are
behind the slits on
the side of a fish's
head.

ACTIVITY

What helps a fish live in water?

1. Look at a fish with a hand lens.

2. What do the scales look like?
   What do they feel like?
   How do they protect the fish?

3. Can you find the gills?
   What do they look like?
   What do they feel like?
   How do they help the fish to breathe?

4. How many fins
   does it have?
   How do fins
   help it move?

5. Draw a picture
   of the fish.

PEOPLE AND SCIENCE

This person is a doctor for animals. An animal doctor takes care of many kinds of animals. The animals cannot tell the doctor how they feel. The doctor uses her eyes, ears, nose, and hands to check each animal. The doctor knows what each kind of animal needs to stay healthy.

Main Ideas

- Mammals are warm-blooded animals with hair.

- Birds are warm-blooded animals with feathers.

- Fish are cold-blooded animals that live in the water and have gills.

Science Words

Match each word with the same part in the picture.

mammal	fins	wings
feathers	bird	hair
beak	fish	gills
scales		

REVIEW

Questions

1. Mammals breathe air with their _____ .

2. Fish use their _____ to breathe.

3. What body parts help fish live in water?

4. Look at these pictures. Group the animals—mammal, bird, or fish.

5. Which ones are warm-blooded?

CHAPTER 2

MORE ANIMAL GROUPS

1.

REPTILES

Snakes, lizards, and turtles are **reptiles.** Reptiles are covered with scales. Their skin may feel rough and dry.

All reptiles are cold-blooded. When the air is warm, a reptile feels warm. When it is cold outside, their bodies are cold. If a reptile is cold, it cannot move fast.

Most reptiles lay eggs. The eggs are laid on the land. The babies hatch from the eggs. The parents do not feed their babies. How do you think the babies get food?

Many reptiles can swim. They do not breathe under water. Reptiles breathe air with their lungs.

Many reptiles are the same color as the things around them. Can you find the reptile in the picture? How does its color help it?

Some reptiles use their tongues to help them smell.

Some reptiles shed their skin as they grow. This lizard is shedding its skin. There is new skin under the old skin.

Reptiles live in forests, jungles, and
deserts. What reptiles live near you?

Crocodiles and alligators are reptiles, too.
How are these two reptiles the same?

ACTIVITY

Can you group reptiles?

1. Look at some books about animals.

2. Which animals are reptiles?

3. Make a list of all the reptiles you find in the books.

4. Put the reptiles into groups. What are the names of your groups?

2.

AMPHIBIANS

These salamanders are **amphibians.**
Amphibians are cold-blooded animals. Most
of them have wet, smooth skin. They have
no hair or scales. How are amphibians
different from reptiles?

A frog is an amphibian.

A toad is an amphibian, too. When amphibians are young, they live underwater. When they are older, they live on the land.

Amphibians lay eggs in water.
Many eggs are laid together.

The eggs do not have hard shells.
The babies hatch from the eggs.

The babies breathe and grow in water. They have gills like fish. They can swim like fish. Baby amphibians find food under water.

Baby amphibians are **tadpoles**. A tadpole begins to look more like its parents as it gets older. It grows legs and lungs. Then it can live on the land.

ACTIVITY

What amphibians live near you?

1. Get a book about amphibians.

2. Look at its pictures and maps.

3. Find the amphibians that live near you.

4. Draw their pictures.

3.

INSECTS

A grasshopper is an **insect.** All insects have six legs. Insects have three body parts. Can you see the body parts and legs? Most insects have four wings. The grasshopper's wings are folded down. Can you name other insects?

This is an adult praying mantis. It is an insect.

All insects lay eggs. Some baby insects look like their parents. These praying mantis babies just hatched. They are very tiny. But they look like their parents. Each baby has six legs. Each baby has three body parts.

Some baby insects do not look like their parents. This is an egg laid by a butterfly.

The baby that hatches is called a **caterpillar.** The caterpillar eats and grows.

Soon it stops eating. It stays very still. It changes into a **pupa.**

Inside the pupa the insect is changing. A grown-up butterfly climbs out.

ACTIVITY

Make a home for insects.

1. Look for insects on plants.

2. Put an insect in a jar.

3. Add some leaves from the same plants.

4. Cover the jar with a net.

5. Watch your insect. How does it move? How does it eat?

6. Draw your insect. Show its body parts.

4.

ANIMALS OF LONG AGO

The earth is always changing. Long
ago it looked very different. The animals
of long ago were different, too. How do
we know? People find **fossils.** Fossils are
made of rock. Some fossils have
footprints in them. Other fossils show
the shapes of animals.

These bones are fossils, too. The bones
came from an animal of long ago. They
took many years to change to rock. The
bones are very big. They came from an
animal that is no longer alive. This
animal is called a **dinosaur.**

Dinosaurs were reptiles. They laid
eggs on land. Some dinosaurs had scales
on their skin. Some dinosaurs were as
small as a chicken. These two animals
were very big. They could look over the
top of a house!

Not all animals of long ago were reptiles. What kinds of animals were these? One looks like a reptile. But it had feathers and could fly.

Mammals also lived long ago. Which animal looks like a mammal? What animal of today does it look like?

PEOPLE AND SCIENCE

These people are looking for dinosaur fossils. They dig them out of the ground. They hope to find all the bones of the animal.

The bones are taken to a museum. The bones are like the pieces of a puzzle. They fit together to make a skeleton. People visit the museum to see the skeleton.

Main Ideas

- Reptiles are cold-blooded animals with dry, rough skin.

- Amphibians are animals that live in water and on land.

- Insects are animals that have six legs and three body parts.

- Some animals that lived long ago do not live today.

Science Words

Match each word with a picture.

amphibian	**fossil**	**reptile**
dinosaur	**insect**	**tadpole**

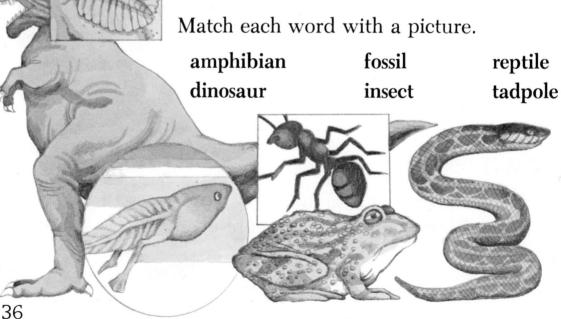

Questions

1. How are these two animals the same?

2. How are they different?

3. How does this insect grow up? Tell the right order of the pictures.

Science Project

Visit a library or museum. Find out about the earth long ago. What did plants look like? What kinds of animals were there? Draw how they looked.

37

CHAPTER 3

WEATHER

1.

WHAT MAKES WEATHER?

Look out your window. What does
the sky look like? Is the wind blowing?
Is the air cool or warm? What is the
weather? Sun, air, and water make
the weather.

Look at the picture. The sun is shining. It is a sunny day. Is the weather warm or cold? How can you tell?

Moving air is called **wind.** The wind can move things. Look at the sails on the boat. The wind pushes them. Which direction is the wind coming from?

The sky is covered with **clouds.**
The sun is behind the clouds. The sun
cannot be seen. It is a cloudy day.

Have you ever walked through a
cloud? A low cloud near the ground is
called **fog.** What kind of day is it in
this picture?

Rain falls from the clouds. What does it look like outside on a rainy day?

Snow falls from clouds. It piles up on the ground. It is cold outside. When the weather gets warm, snow melts. What can you do with snow?

ACTIVITY

What will the weather be?

1. Draw some weather pictures.

2. What is the weather today?
   Put a picture of it on a chart.

3. What do you think the weather will
   be tomorrow? Put a picture of it on
   the chart. Now, each day on the
   chart will have 2 pictures.

4. What does it mean if the pictures
   match?

2.

HOT AND COLD

The air around us can be warm or
cool. In the morning the air feels cool.
The children wear jackets and sweaters.

In the afternoon the air is warmer.
How are the children dressed? What
made the air warm? The sun warms the
air during the day.

The sun **heats** the earth. It heats the land and water. It makes the air warm, too. At night we have no sunlight. The night air is cooler.

Temperature tells you how hot or cold something is. It is important to know the air temperature. Do you know why?

We measure temperature by using a **thermometer.** The liquid inside the thermometer moves up as the temperature gets hotter. Look at this thermometer. Is the temperature hot or cold?

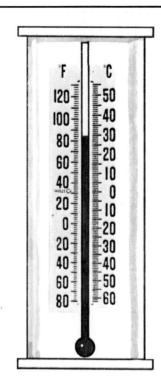

The liquid inside the thermometer moves down as the temperature gets colder. Look at this thermometer. Is the temperature hot or cold?

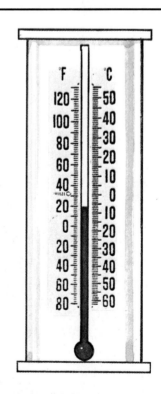

ACTIVITY

What is the temperature?

1. Put a thermometer in cold water.

2. Put a thermometer in warm water.

3. Watch the liquid inside the thermometer.

4. What happens to the liquid when it gets warmer?

5. What happens to the liquid when it gets colder?

3.

WEATHER CHANGES

Weather changes from day to day. It changes from **season** to season. It is hot in the summer and cold in the winter. It may be dry in the summer and rainy in the winter.

Weather changes make living things change. How does the weather change the things you do?

Plants change as the weather and seasons change. The leaves of some plants change color in the fall.

Some plants lose their leaves in the winter.

The weather changes in the spring. The air gets warmer. Many plants grow flowers in the spring.

49

Animals do different things as the weather changes. In the winter some animals move to places where the weather is warm.

Many animals have babies in the spring.

In the winter, the daytime is shorter than the nighttime. It gets dark in the afternoon. In places where it is cold, people wear heavy clothes to keep themselves warm.

In the summer, daytime is longer than nighttime. The sun feels very hot. People wear light clothes to keep themselves cool in the summer.

ACTIVITY

In what weather do seeds grow best?

1. Put a wet towel in each of 2 dishes.

2. Put some seeds on the towels.

3. Cover the seeds with another wet towel.

4. Put one dish in a warm place. Put the other in a cool place.

5. Keep the towels wet.

6. Check your seeds every day.

7. Which seeds grew faster? Why?

8. What kind of weather is best for your seeds?

PEOPLE AND SCIENCE

Weather is important to people who fly airplanes. An air traffic controller tells the pilots if it is safe to fly.

If there is too much rain, fog, snow, or wind, the planes may not fly. The air traffic controllers have machines that help them. Their machines tell them if storms are near.

CHAPTER

Main Ideas

- Sun, air, and water make the weather.

- Thermometers are used to measure how hot or cold it is.

- Weather changes from season to season.

- People, plants, and animals change as the weather changes.

Science Words

Pick the word that fits best in each blank.

snow	seasons	rain
wind	thermometer	fog

1. A _____ tells us the temperature.

2. _____ is moving air.

3. Low clouds near the ground are _____.

4. Summer and winter are two _____.

5. _____ falls from clouds as white flakes.

6. _____ is water that falls from clouds.

REVIEW

Questions

1. Look at the picture. What is the weather?

2. How will the tree look in winter?

3. Which thermometer shows a hot day? Which thermometer shows a cold day?

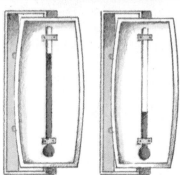

Science Project

Find out how much rain falls. Get a clear jar with straight sides. Put a ruler on the side. Put the jar outside. After it rains, look at the jar. How much water is in it?

CHAPTER 4

WATER IN THE AIR

1.

WHAT IS AIR?

Air is all around our earth. We call
air a **gas.** It is the gas that we breathe.
We need air to live. We cannot see,
smell, or taste air. What can you see
in the air?

57

The air around the earth is always moving. Wind is air on the move. Wind can push things. Sometimes the wind is very strong. Sometimes the wind is weak. Watch the clouds on a windy day. The wind pushes them.

We feel changes in the air. The air can feel warm or cool. The air can feel clean and dry.

Sometimes the air feels wet. It makes our skin feel sticky. Wet air makes you feel hot. How does the air feel here? Is it warm or cool? Is it dry or wet?

ACTIVITY

How strong is the wind?

1. Look outside.

2. What is the wind able to move?

3. Look at the chart.
   What kind of wind is it today?

4. Check the wind later in the day.
   Did it change?

WIND FORCE	
Name of wind	What you can see
Calm	Leaves on trees are still. Smoke rises straight up.
Gentle Breeze	Leaves moving. Wind moves flag a bit.
Strong Breeze	Large branches are moving. Umbrellas are hard to use.
Gale	Twigs snap off trees. Hard to walk.

2.

WHERE DOES WATER GO?

Oh! What fun! It has just rained. The children are playing in rain puddles. The sun is shining on the puddles. Soon the puddles will be gone. Where does the water go?

Heat helps make water go into the air. Where did heat come from to dry up the ground?

The wind helps make the water go into the air, too. What is helping to dry these clothes? Where does the water go?

Water that is in the air is called **water vapor.** What is making water change to water vapor? Water goes into the air in many places. Can you name some of these places?

ACTIVITY

Where does water go?

1. Fill some jars with water.

2. Mark the top of the water.

3. Put the jars around your room.

4. What do you think will happen?

5. Check the jars in 2 or 3 days.
   What happened to the water?

6. Did the water change by the same
   amount in every jar?

3.

WHERE DOES WATER COME FROM?

We cannot always see water vapor.
But it is always in the air. Water vapor
can change back into water. Have you
ever seen water on plants in the
morning? The cool night air changed
water vapor back to water. This water
is called **dew**.

The sun shines on the plants. Soon the
dew is gone. Where did it go?

The air is very cold high above the earth. In cold air the water vapor changes into drops of water. The drops of water and dust in the air make **clouds**.

The cloud drops get bigger and bigger. The drops of water get too heavy to float in the air. Soon they begin to fall. Then we have rain.

1. Heat from the sun changes water into water vapor.

2. The water vapor goes into the air.

3. Cold air changes water vapor back to drops of water.

4. The drops of water mix with dust to make clouds. What do you think will happen next?

ACTIVITY

When does water vapor turn into water?

1. Fill a glass with ice water.

2. Fill another glass with warm water.

3. Let the glasses sit for a few minutes.

4. Feel the outside of both glasses.
   Do they feel the same?

5. Explain what happened.

4.

CLOUDS

There are many kinds of clouds. Their shapes, colors, and sizes are different. Clouds help us know what the weather will be.

These clouds are very big. They cover the sky. They look dark because they are made of many water drops. They are rain clouds.

These clouds look like cotton balls. They seem to be always moving and changing shape. The sun peeks through these clouds.

When the clouds look like this, the weather will be good. How are these clouds different from rain clouds?

You can see these clouds on a clear, sunny day. They are the highest clouds in the sky. They are so thin you can almost see through them. They look like feathers in the sky.

The air is very cold high in the sky. These clouds are made of bits of ice.

ACTIVITY

What kinds of clouds can you see?

1. Look at the different kinds of clouds.
   How are the clouds alike?
   How are they different?

2. Look out the window.
   Do you see clouds that look like any
   of the clouds in the pictures?

3. Draw a picture of the clouds and the
   weather each day for one week.

In some places the air gets very cold. The water vapor changes to ice. Ice in the air is called snow. It's fun to ski on snow.

Sometimes not enough snow falls. This person works a snow maker. The snow maker shoots tiny drops of water into the cold air. The water freezes and falls on the ground. The snow maker works all night to cover the ski slope.

CHAPTER

Main Ideas

- The earth's air is always changing.

- Heat and wind change water into water vapor.

- Cool air changes water vapor into water.

- There are many kinds of clouds.

Science Words

The Science Words letters are mixed up.
Write the words the right way.

1. Water in the air is *trwea pavro.*

2. Shapes in the sky made of water drops are called *olcdus.*

3. Drops of water on the ground in the morning is called *ewd.*

4. The sun's *teha* makes the water in puddles go into the air.

REVIEW

Questions

1. Which set of clothes will dry fastest?

2. What are clouds made of?

3. What makes clouds move in the sky?

4. When you see feathery clouds you will have _____ weather.

5. What moves in the air that you like?

6. What moves in the air that is bad for you?

CHAPTER 5

SOUND

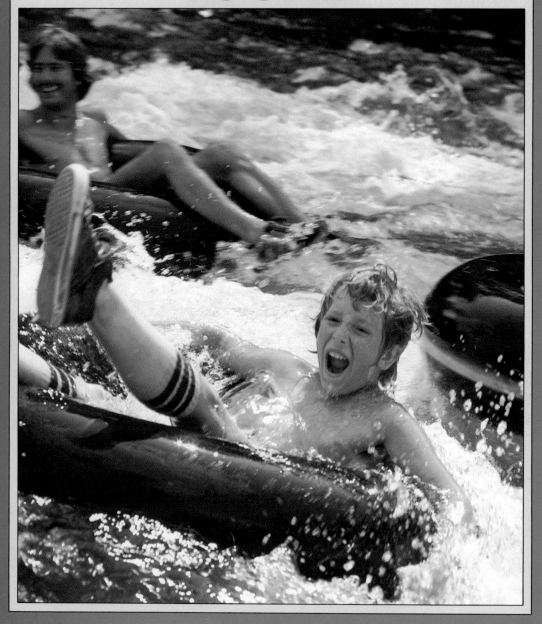

1.

WHAT IS SOUND?

A **sound** is made when something moves back and forth very fast. When things move back and forth, they **vibrate.** What things in the picture are making sounds? The strings of the harp are vibrating. They move back and forth. They are making sounds.

77

This girl tapped a glass filled with water. Her tap made the glass, the water, and the air inside the glass move fast. This made a sound.

Tap on a desk. Do you see anything move back and forth fast? Sometimes things vibrate so fast that you do not see them move.

Sounds are made when air moves back and forth fast. Blowing into a bottle makes the air inside move fast. The air vibrates. This makes a sound.

Blowing into a whistle makes sounds. The air in the whistle vibrates. How are these children making sounds?

ACTIVITY

Can a ruler vibrate?

1. Press a ruler down on a desk.

2. Gently snap the end that sticks out.

3. Which picture did it look like?
   What did you hear?

4. Snap the ruler harder.

5. Which picture did it look like?
   How did the sound change?

2.

SOUNDS ARE DIFFERENT

We know things by their sounds. Sounds are different from each other. Close your eyes. What sounds do you hear? How are they different?

Some sounds are **high.** A whistle is a high sound. What high sounds do you know? Some sounds are **low.** A cow's moo is a low sound. Can you think of other low sounds?

Some sounds are **loud.** A fire truck's horn makes a loud sound. Why is it loud? What loud sounds do you know?

Some sounds are **soft.** A whisper is a soft sound. What soft sounds do you know?

Some sounds are too loud for us. Very loud sounds can hurt our ears. What is making loud sounds here?

This worker is taking care of his ears. He wears a headset to block the loud noises. What other loud noises hurt your ears?

ACTIVITY

What makes sounds different?

1. Put a rubber band around a ruler.

2. Put a pencil under the rubber band.

3. How can you make a loud sound?

4. How can you make a soft sound?

5. Move the pencil.

6. How can you make a high sound?

7. How can you make a low sound?

3.

SOUND MAKERS

The world is full of sounds. Animals make sounds. A cricket vibrates its legs to make sound. Machines make sounds.

What sounds does the weather make? Raindrops, thunder, and wind make the sounds of a storm. Every time a sound is made, something vibrates.

Some sounds make music. What things are these children using to make music? What part of each music maker vibrates?

Animals make many sounds. These animals make sounds with their **throats.** The air in the throat vibrates. The vibrating air makes a sound.

What kind of sound does a kitten make? Is it high or low?

Kittens and lions use their throats to make sounds. In what other ways do animals make sounds? How do you make sounds?

ACTIVITY

How can you make sounds?

1. Put your hand on your throat.

2. Hum softly.

3. What do you feel? What part of you vibrates?

4. Hum louder. How does it feel now?

5. Make other sounds with your throat.

6. List words for the sounds you made.

4.

SOUNDS MOVE

Sounds can move. They move to our
ears. Sounds move through the air.
There is air between the teacher and the
students. The teacher's throat makes the
air vibrate. When air vibrates, small
parts inside your ear vibrate, too. Then
you hear the sound.

Sounds also move through objects.
One child makes a sound. The sound
makes the string move. The string
vibrates. Sounds move through the
string. The sound moves into the other
child's ear.

Sounds move through some things
better than others. A boy is walking on
a wood floor. The sound is loud.

The other boy is walking on a carpet.
The sound is soft. Does sound move
better through wood or carpet? How do
you know?

ACTIVITY

Does sound travel better through cotton or air?

1. Fill a bag with cotton.

2. Put the bag next to your ear.
   Listen to a clock.

3. Can you hear the clock tick?

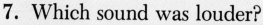

4. Fill a bag with air.

5. Listen to the clock through the bag.

6. How does the tick sound now?

7. Which sound was louder?

8. Did sound travel better through cotton or air?

PEOPLE AND SCIENCE

Did you ever look inside a piano? Could you build one? That's what people in a piano factory do.

This person pulls steel wires across a metal frame. The wires are called strings. Piano strings must be very tight. The strings vibrate to make music.

Main Ideas

- A sound is made when something moves back and forth very fast.

- Sounds are high, low, loud, or soft.

- Sounds travel to our ears.

- Sounds travel better through some things than through others.

Science Words

Match each word with a picture.
Tell how each thing sounds.

high   low   loud   soft

REVIEW

Questions

1. Look at the picture. What will happen to the beans when the drum is hit?

2. What part of the drum vibrates?

3. What sounds can hurt your ears?

   high   **low**   **loud**   **soft**

4. How does your throat make sounds?

5. Why do some schools have carpet on the floors?

Science Project

Collect things that you can use in a band. Make up a beat for your instruments. Teach it to your friends.

CHAPTER 6

LIGHT

1.

LIGHT MAKERS

Light comes from many things. The sun makes its own light. It is a giant ball of fire. The sun is a **bright** light.

A candle makes its own light. It is not a bright light. The candle's flame is small. It is a **dim** light.

We do not see the sun's light at night.
How do we see at night? People make
light bulbs to help us see. Light bulbs
make their own light.

Where are light bulbs used? Are they
bright or dim? What other things make
their own light?

We cannot see without light. Places
without light are very dark. You can't
see in the dark. A closet is dark. What
helps you see in a closet?

Do you like a light in your room
at night? How does it help you?

ACTIVITY

Do you need light to see things?

1. Tape a crayon inside one end of a shoebox.

2. Cut a small hole in the other end of the box. Put the lid on.

3. Roll up a piece of black paper. Fit it in the hole.

4. Look into the box. What do you see?

5. Lift the lid of the box just a little. What do you see now?

6. What do you need to see?

2.

HOW LIGHT MOVES

Light moves in a straight line. The sun's light moves through space. It hits only part of the earth at one time. The earth turns. Each part has light and then darkness.

When the sun's light shines, it is **day.** What do you do in the day? It is dark on the other side of the earth. We call this **night.** What do you do at night?

When light moves, it hits things
in its path. Light bounces off things
in straight lines.

The flashlight shines into a mirror.
The light **reflects** off the mirror. It
reflects in a straight line. Where does
the light move next?

Most things do not make their own light. We see these things when light reflects off of them. The light reflects into our eyes.

How do we see the actress? A light bulb makes the light. The light bounces off the actress. Then the light moves to our eyes.

ACTIVITY

How can light go around corners?

1. Stand around a corner from a friend.

2. Turn on a flashlight.

3. Does the light hit your friend?

4. Have another student hold a mirror.

5. Aim the light at the mirror.

6. Can the mirror make the light hit your friend?

7. What did the mirror do to the light?

3.

LETTING LIGHT IN

These children are at the zoo. The clean glass makes it easy for them to see. Light moves through the glass easily. What do the children see through the glass? Where else do people use glass?

Sometimes the glass gets dirty. Then it is hard for light to move through it. The dirt on the glass blocks the light. Where else does glass get dirty?

What is over the ruler, scissors, and pencil? All light passes through them. We can see through these things.

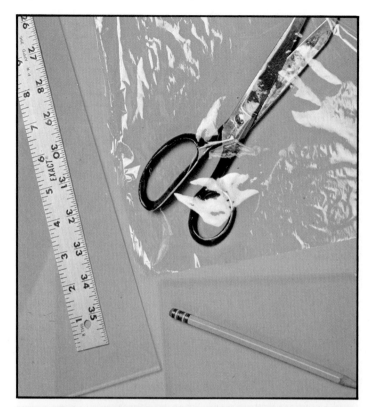

The ruler, scissors, and pencil are now harder to see. What is over them? Only some light passes through these things. We cannot see clearly through these things.

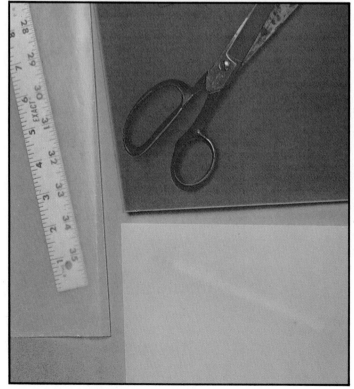

No light passes through these things. We cannot see through these things.

What things in this room let all light through? Which things let some light through? Which things let no light through?

ACTIVITY

How much light passes through?

1. Copy the chart from the picture.

2. Hold some objects up to a light.
   Look at the objects carefully.

3. How much light passes through?

4. Write your answers on the chart.

4.

SHADOWS

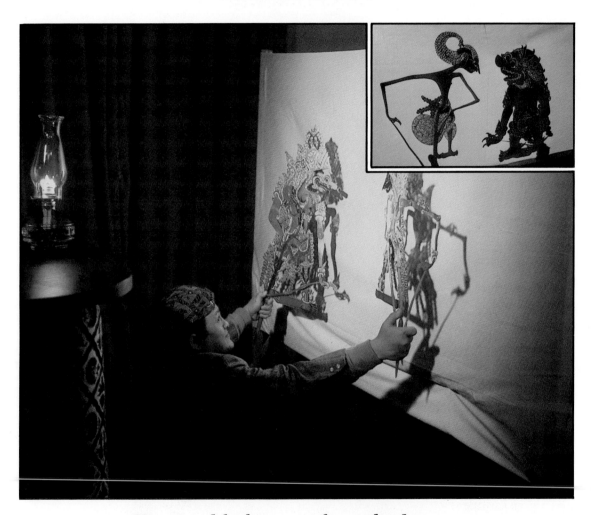

You need light to make a **shadow.**
The light can come from the sun. Or, it
can come from a lamp.

Something must block the light to
make a shadow. Look at the shadows in
the picture. What is blocking the light?

You make a shadow. Your body blocks
the light. Shadows can change. You can
move the light. The size of the shadow
will change. You can make your shadow
big or small.

Look at your shadow. Sometimes it looks the same as you. You can turn your body. The shape of your shadow will change. What kinds of shadow shapes can you make?

111

ACTIVITY

Do sun shadows change?

1. Go outside in the morning sunlight.

2. Ask a friend to trace your shadow.

3. Measure the length of your shadow. How many centimeters long is it?

4. Trace your shadow at noontime.

5. How long is it now?

6. How did your shadow change?

7. What made it change?

PEOPLE AND SCIENCE

People come to see plays here. This is a theater. Light is used in many ways here. This person moves the lights. A spotlight hits only one actor at a time.

The lights can be bright or dim. The color of the lights can change. Lights help make the play happy, sad, or scary.

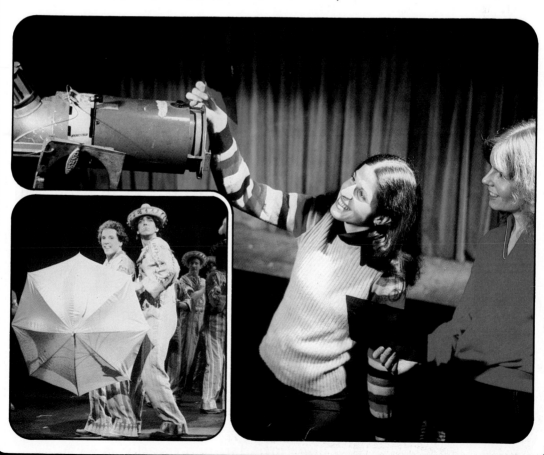

Main Ideas

- We need light to see.

- Some things give off their own light.

- Light travels in a straight line.

- Light bounces off many things.

- Light travels through some things better than through others.

- Things that block the light make shadows.

Science Words

Pick the word that fits best in each blank.

day	**bright**	**light**
dim	**night**	

1. A _____ light makes a dim light.

2. We cannot see without _____.

3. The sun is a _____ light in the _____.

4. The light from a candle is very _____.

REVIEW

Questions

1. What things give off their own light?

2. Where is the boy's shadow?

3. Look at the objects. Which ones let all light pass through?

4. Which objects let a little light pass through?

5. Which ones let no light pass through?

6. Which ones will make the darkest shadows?

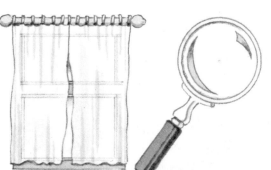

CHAPTER 7

FORCE

1.

WHAT IS A FORCE?

A weightlifter **pulls** the bar up from
the floor. Then he **pushes** it above his
head. People, machines, and animals
can push or pull.

A motorboat is a machine. It can pull
a person on waterskis. The boat pulls
the man over the water.

A **force** is a push or a pull. What kind of force do you see here? What does this machine do?

The boy is using a force. What kind of force is it? What forces do you use when you play?

A **strong** force is a big push or pull.
This big engine is pulling the train. It
pulls with a strong force.

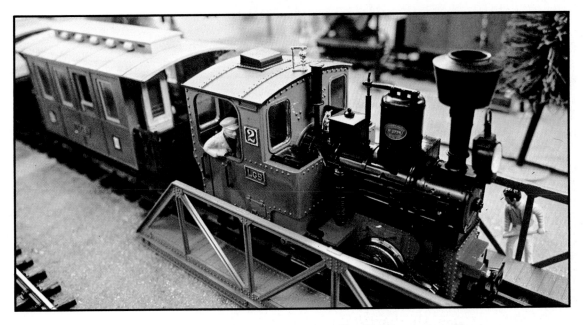

A **weak** force is a small push or pull.
This is a toy engine. It pulls with a
weak force.

ACTIVITY

Is it a push or a pull?

1. Move your chair under your desk.

2. Did you use a push or a pull?

3. Make a chart of things to move.

4. Move each thing.

5. Did you move it with a push or a pull?

6. Mark your answers on your chart.

7. Which force did you use the most?

What we moved	Push	Pull

2.

FORCES MOVE THINGS

An elephant is hard to move. Many students are pulling. They use a lot of force. Will they be able to move the elephant?

Heavy things are hard to move.
Which wagon has a heavier load? How
do you know? A strong force is needed
to move heavy things.

A weak force is needed to move **light**
things. Which wagon do you think needs
less force to be moved?

The boy moves the weight up and
down. When does he use more force?

Moving up a hill is harder than going
down. Which child uses more force?

ACTIVITY

More force or less force?

1. Make an uphill ramp.

2. Tie a string to a paper cup and an object.

3. Put paper clips in the cup one by one.

4. How many paper clips does it take to move the object uphill?

5. What kind of force moved it?

6. Would it take more or fewer paper clips to move the object downhill?

3.

FORCES CHANGE THINGS

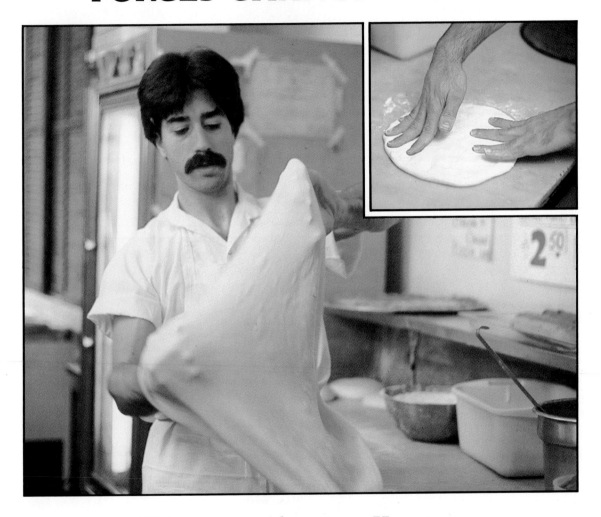

This person makes pizza. He gets a round piece of dough. The dough is pushed. It makes a flat **shape**.

Then the dough is pulled. Pulling changes the **size**. It gets bigger. Now it's ready for sauce and cheese!

Forces change the
speed of things. How
can the girl go **faster**?
She pushes harder.
She uses more force.
The bike moves faster.

Forces can make things go **slower**.
The dog wants to go fast. The boy pulls
hard. He uses a strong force. The force
makes the dog go slower.

Forces can change the **direction** things move in. The girl changed the direction of the ball. What kind of force did she use?

The truck turns a corner. It pulls its load along. It uses a force to change direction.

ACTIVITY

What changes do forces make?

1. Measure a clay ball with your ruler.

2. Make the clay ball flat.

3. How long is it?

4. Make the clay ball longer.

5. How long is it now?

6. What forces did you use to change the clay?

4.

HOW MUCH FORCE?

These people are pushing carts in a store. The cart with more things is heavier. It takes more force to move it. Which person must push harder?

129

The horses are pulling loads of hay.
How many horses are pulling each load?
Two horses have more force than one.
Which load needs more force to move?
Which load **weighs** more?

The boy wants to lift the toys. He is pushing down on the seesaw. The toys are too heavy for him to move. They weigh a lot. What can he do?

Now a friend helps him. The force of two children can move the toys.

ACTIVITY

**How many paper clips are needed
to lift things?**

1. Set up a balance.

2. Put a crayon on side A.

3. Put paper clips on side B until the
   sides are straight across from each other.

4. How many paper clips weigh the
   same as a crayon?

5. Weigh other things in your desk.

6. Order the things from lightest to heaviest.

Visiting a gym is fun. It is also good for your health. A gym has many machines for people to move. The machines make you work hard.

You use the force of your body to move the machine parts. You can pull or push on a bar with weights. Pushing and pulling make your muscles stronger.

Main Ideas

- A force is a push or a pull.

- A force is needed to move things.

- A force can change the size and shape of things.

- A force can change the speed or direction of things.

- A force can be measured.

Science Words

Tell about the pictures.
Use these words.

pull	weak	push
strong	heavy	light

REVIEW

Questions

Find the picture that answers each question.

1. Which shows a change in direction?

2. Which shows a tool used to tell how much something weighs?

3. Which shows a change in the speed of a moving thing?

4. Which shows a weak force moving a light load?

Science Project

People, animals, and machines lift things. Find pictures of things being lifted. Is a push or a pull used to lift?

CHAPTER 8

MOVING

1.

MOVING SLOWLY

Did you ever hike on a mountain path? Hiking can be hard work. These hikers are moving slowly. Their boots rub against the stones. The stones are slowing them down.

Moving on something **rough** slows
things. What is each child moving on?
Which child moves slower? Which child
needs more force to move?

When something moves, it rubs against other things. Rubbing slows moving things down. The truck's wheels rub against the dirt. The dirt is very rough. The dirt makes the truck move slowly.

You can throw a ball through the air. The ball rubs against the air. Air slows things down.

ACTIVITY

What surfaces slow things down?

1. Push a block of wood over sandpaper.

2. Is it a rough or a smooth surface?

3. Push the wood across a tile floor, a carpet, a gym floor, and a sidewalk.

4. Which ones are rough surfaces?

5. On which ones does the wood move slowly? On which ones does the wood move quickly?

2.

SLOWING THINGS DOWN

Ice is very **smooth.** The ice skates move easily. Sometimes the skates move too fast.

This girl wants to stop. She turns and points her toe on the ice. Look at the toe of the skate. How will it help her stop?

141

Snow and ice make roads slippery.
Sand helps make the road rough. Sand
helps slow cars down.

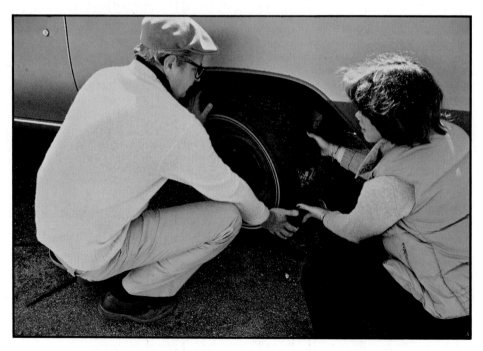

Snow tires are very rough. They help
cars slow down without slipping.

Some shoes are made to help keep you from slipping. Their bottoms are rough. They help you slow down.

Look at the bottoms of these shoes. Which ones are for walking on snow? Which ones can grip wet grass? The white ones are for tennis. They let you slide a little bit.

143

ACTIVITY

Why do things move differently?

1. Make a low ramp.

2. Put a toy car at the top and let it go.

3. Measure how far the car traveled.

4. Cover the ramp with sandpaper.

5. Put the car at the top and let it go.

6. Measure how far the car traveled.

7. When did the car travel farther?

8. Which surface slowed the car down?

3.

MOVING THINGS EASILY

Sometimes we want things to move faster and more easily. Things move faster and more easily on smooth things. The water slide is smooth. Are these people moving fast or slowly? Is it easy or hard for them to move? What would happen if the water were turned off?

The girl is putting oil on her toy. The oil makes the toy's parts slippery. The parts move smoothly against each other. This makes the toy move easily. What other things need oil to move easily?

Many things have wheels. Wheels let objects roll. Rolling is easier than sliding.

Wheels make it easier to move things. You can move heavy things with wheels.

ACTIVITY

Why do things move more easily?

1. Rub two fingers together.

2. Put some oil on your fingers.

3. Rub your fingers together again.

4. Does the oil feel smooth or rough?

5. Make a chart.

6. Do the same thing with glue, powder, and sand.

7. What happens? Write it down.

8. Which ones help things move more easily?

PEOPLE AND SCIENCE

Moving fast can save people's lives. Firefighters must get to a fire as fast as they can.

Many firehouses have a big pole. It joins the upstairs with the garage. The firefighters slide down the pole. The pole is very smooth. Sliding down is faster than walking on the stairs.

Main Ideas

- When something moves it rubs against other things.

- Moving over rough places slows things down.

- Smooth places make things move easily.

- Wheels and oil make things move easily.

Science Words

Look at the pictures.
Which show **rough** surfaces?
Which show **smooth** surfaces?

REVIEW

Questions

Answer the questions from the picture.

1. What makes the sled move easily?

2. What will make the sled slow down?

3. How could you stop the sled?

4. Will it take more force to go up or down the hill?

5. Will a sled go fast on a grassy hill?

6. How would a sled with wheels work on a grassy hill?

CHAPTER 9

OCEANS
AND BEACHES

1.

WHERE ARE OCEANS AND BEACHES?

The people are looking at a globe. A globe is a map of the **earth**. The blue parts on the globe show water. There is more water than land on earth. Some of the water is in **lakes** and **rivers**. Most of the water is in **oceans**.

Is there a **beach** near your home?
A beach is where the water meets the
land. The boy is standing on the beach.
Lakes, rivers, and oceans have beaches.

154

There are many kinds of beaches. They are made of different things.

There are rocky beaches.

There are sandy beaches.

Some beaches are made out of crushed shells.

ACTIVITY

Is there more land or water on earth?

1. Get a globe.

2. Cover all the water with red squares. How many squares did you use?

3. Cover all the land with green squares. How many squares cover the land?

4. Is there more land or water on the earth? How do you know?

2.

HOW ARE BEACHES MADE?

There is sand in this lake. Water pushes the sand up onto the land. This makes a beach.

Shells are parts of sea animals. The shells are pushed onto the land by the water. This makes a beach.

Water is rubbing sand against these rocks. When this happens, the rocks will slowly break into sand.

The sand at this beach was made a long time ago from very large rocks. What helped change the rocks into the sand?

The wind blows sand against the rocks. The wind and sand scrape the rocks. Scraping changes the rocks into sand. Wind and water work together to help make beaches.

ACTIVITY

Where does sand come from?

1. Rub two rocks together over a paper towel.

2. What falls into the towel?

3. What did you do to break the rock apart?

4. What breaks rocks apart at a beach to make sand?

3.

OCEAN WATER

These people are sailing on the ocean. Water in the ocean is different from the water we drink. We cannot drink the ocean water. Ocean water is **salty**.

People drink **fresh** water. Fresh water does not have much salt in it. The water in lakes and rivers is fresh. The water we drink comes from lakes and rivers.

The water in
the ocean is always
moving. **Waves** are
made when the
wind moves water.

At certain times
each day, the ocean
water moves away
from the beach.
When this happens,
it is called **low tide**.

At other times
each day, the ocean
water moves closer
to the beach. When
this happens, it is
called **high tide**.

Small animals and plants live on the
beach. Walk on a beach at low tide.
Look near the rocks and puddles. You
can find ocean animals and plants. The
animals hold tight to the rocks.

ACTIVITY

How is salt water different from fresh water?

1. Get a cup of fresh water.

2. Tell how it tastes, feels, looks, and smells.

3. Mix a spoonful of salt in the water.

4. Tell how it tastes, feels, looks, and smells now.

5. How is salt water different from fresh water?

4.

COLD AND WARM WATER

These people are divers. The ocean is
not **deep** here. It is **shallow**. The divers
are near the land. The water is warm
where the divers are. How do you know?

It is warm here. The sun is shining on
the top of the ocean. The water is warm.

Ocean water far from land is very
deep. The sun does not shine deep down
in the ocean. The water is very cold
here. The diver wears a suit to keep warm.

In warm parts of the world, the ocean
water is warm. In cold parts of the
world, the ocean water is cold.

Is the ocean water in the picture
warm or cold? How do you know?

ACTIVITY

Does the sun warm the ocean?

1. Fill a bucket with cool water.

2. Take the temperature at the bottom.
   Take the temperature near the top.

3. Are the temperatures the same
   or different?

4. Shine a lamp at the top of the water.
   Leave it on for 30 minutes.

5. Take the temperatures again.

6. Have they changed? Why?

Glass factories use a lot of sand. The sand is mixed with other materials. Then it is heated until it melts. A worker gathers the hot, glowing glass on a pipe.

The glassblower blows air into the pipe. A bubble is blown to make a bowl. Soon the glass cools. The glass bowl is then cut off from the pipe.

Main Ideas

- Most water on earth is found in oceans.

- Beaches are found near water.

- Wind and water help make and change beaches.

- Ocean water can be warm or cold.

Science Words

Match each word to the right number on the picture.

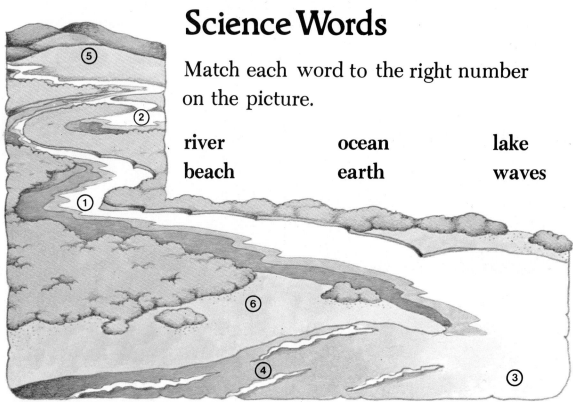

river	ocean	lake
beach	earth	waves

REVIEW

Questions

1. How are these made into beaches?

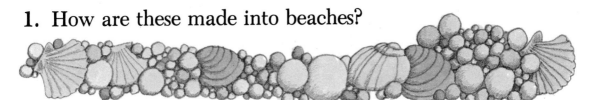

2. _____ pushes water to make waves.

3. How are ocean and lake water different?

4. Match the words with their pictures.

low tide   high tide

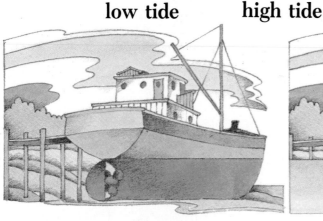

Science Project

Write to someone who lives near a beach. Ask him or her to send you a bit of the beach. Put it in a plastic bag. What is this beach made of?

CHAPTER 10

LIVING IN THE OCEAN

1.

OCEAN PLANTS

Ocean **plants** are different from land plants. Ocean plants can grow in salt water. Some ocean plants float in the water. Some hold tight onto rocks.

The waves push some of the plants onto beaches. These plants are many colors. What color are most land plants?

Most ocean plants need sunlight to
live. They grow only where they can get
sunlight. These plants grow in shallow
water. Can they grow in very deep water?

Many of the plants in the ocean are
very small. They float in the water.
They are so small you cannot see them
without a **microscope**. A microscope
makes things look bigger.

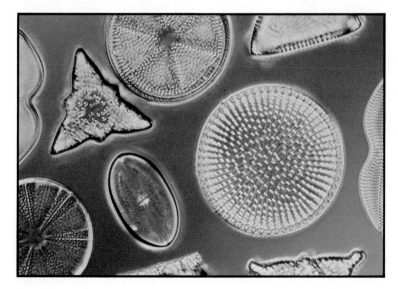

This is what
these small plants
look like through
a microscope.
What shapes do
you see?

ACTIVITY

How are water plants and land plants different?

1. Look at some water plants and land plants.

2. What colors are they?

3. Do they look the same?

4. How does each one feel?

5. Try to stand each one up. Can both do so?

2.

OCEAN ANIMALS

Many kinds of **animals** live in the ocean. They move through the water in different ways. Some have smooth bodies. They swim in the water. Some animals have many legs. They crawl on the bottom.

The blue whale is the biggest animal
on earth. It is as long as three buses. It
lives in the ocean.

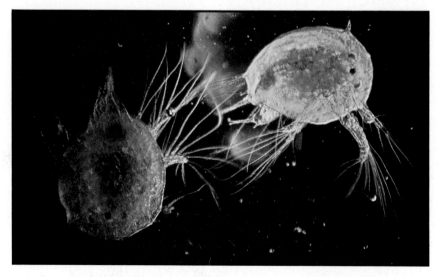

Some of the smallest animals live in
the ocean, too. These tiny animals eat
the small plants.

Look at the two pictures on this page. Are these plants or animals? They are ocean animals that look a lot like plants. They live on rocks at the ocean bottom.

These animals do not move from place to place. How can they find food? They wait for food to come to them. They trap bits of food that float in the water.

179

ACTIVITY

Look at ocean animals.

1. Look at some ocean animals from a fish market. What colors are they?

2. Draw a picture of each one.

3. How do they feel?

4. Look at their body parts. How do they move?

3.

FOOD IN THE OCEAN

All living things need food. Ocean plants use sunlight to make food. Some ocean animals eat plants. Some ocean animals eat other animals. Each living thing is food for another one. This is called a **food chain**.

Here is an ocean food chain.
Let's see how it is put together.

1. Tiny plants make their own food.
   They use sunlight to do it.

2. Tiny animals eat these
   plants for food.

3. Small fish eat the tiny animals.

5. People eat big fish for food.

4. Big fish eat the small fish.

What would happen to the big fish if
all the small fish were gone?   183

ACTIVITY

Are people parts of food chains?

1. Look around a food store.

2. Write down all the foods made from ocean animals.

3. Draw pictures of the animals.

4. Use the pictures to make some ocean food chains. Add people, too.

PEOPLE AND SCIENCE

An aquarium is like a zoo. It is a place where people can see ocean animals. The animals live in tanks of salt water.

This person takes care of the animals. She knows what foods to give to each animal. She keeps the water at the right temperature. She knows what each animal needs to stay healthy.

Main Ideas

- Ocean plants are different from land plants.

- There are many kinds of ocean animals.

- Ocean plants and animals need each other for food.

Science Words

Match each word with a picture.

microscope   **animal**

plant   **food chain**

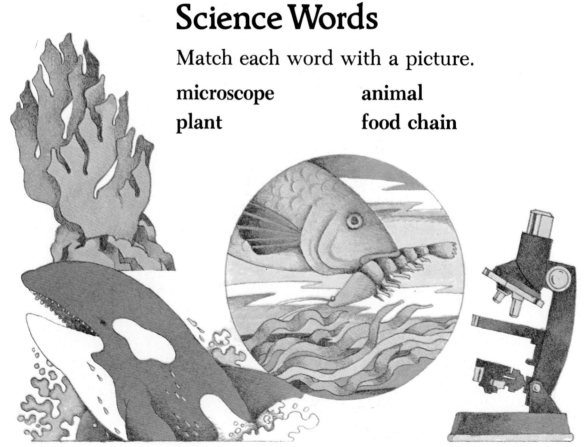

REVIEW

Questions

1. How are these plants different?

2. Name some ocean animals.

3. This food chain is mixed up. What is the right order?

Science Project

Make a shell collection. Group the shells by how they look. Show how the animal uses its shell.

CHAPTER 11

PARTS OF PLANTS

1.

ROOTS

Many plants have **roots**. The root is the part that grows in the ground. Some plants have roots that grow underwater.

Roots are important to plants. The roots of a tree hold tight in the ground. They help the tree stand up. What would happen if the tree had no roots?

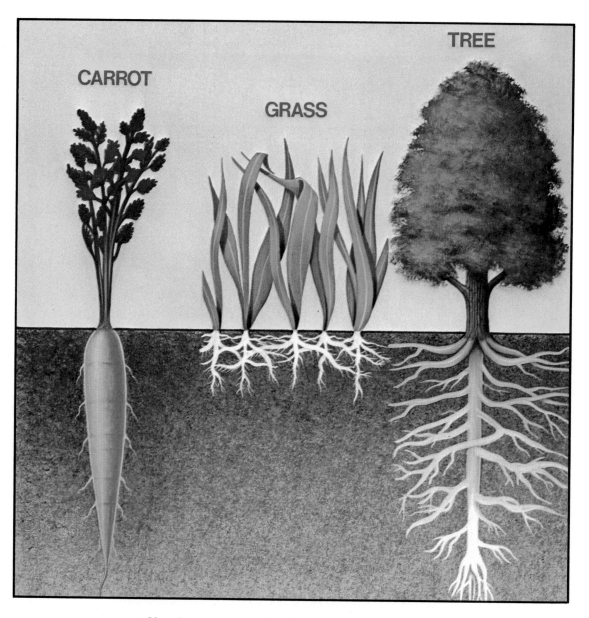

All plants use their roots to get water.
The water travels through the roots to
the rest of the plant.

Roots help the plant in other ways.
Some roots have very thick parts. The
plants store extra food in them.

We eat the roots of many plants. Do
you know the names of these foods?
They are all parts of roots.

ACTIVITY

How does water get into a plant?

1. Grow potato roots like this.

2. Mix red food coloring in the water.

3. Wait a few days. How did the potato change?

4. Cut the potato in half. What do you see?

5. How did it happen?

2.

STEMS

Stems are parts of plants. The stem is between the roots and the leaves. Wildflowers have soft stems. The trunk of a tree is a stem. Tree stems are hard and strong. What do people use tree stems for?

The stem helps the plant in two ways. The stem holds up the plant. It holds the leaves up to the sunlight.

Water moves up the stem. It travels from the roots to the leaves. Food moves down the stem. It comes from the leaves.

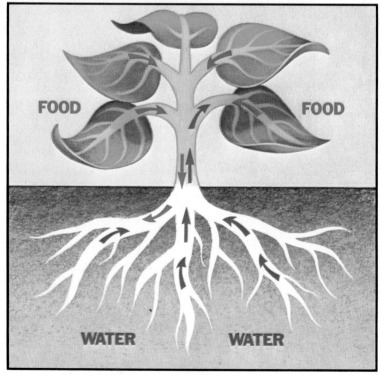

We eat the stems of many plants. Do
you know the names of these foods?
They are all stems of plants. What other
plant stems do you eat?

Stems grow under and above the ground.
The part above the ground is green.

195

ACTIVITY

How does water get to the leaves?

1. Mix red coloring in some water.

2. Put a celery stem in the water.

3. Wait a day.

4. How do the celery leaves look?

5. Cut off a piece of the stem. What do you see?

6. What is one job of the stem?

3.

THE LEAF

Each of these objects is a **leaf**. A leaf
is a part of a plant. The leaf grows on
the stem. How do they look the same?
How do they look different? All leaves
do the same job for plants.

The leaf makes food for the plant. It uses air, water, and sunlight. Water rises from the roots. Air comes in from tiny holes in the leaf.

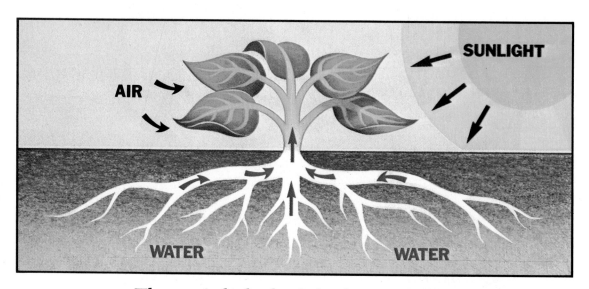

SUNLIGHT

AIR

WATER   WATER

The sun's light hits the leaves. Now the leaf can make food. Only green plants can make food.

Some leaves change color. They turn from green to brown. The brown leaves cannot make food. Soon the leaves drop off the stems. Food is not made now. In spring, new leaves grow. The leaves are green. Can they make food now?

People eat the leaves of many plants. These green leaves are very good for you. Do you know the names of these leaves? Which ones do you eat cooked? Which ones do you eat raw? What other plant leaves do you eat?

ACTIVITY

Does a leaf give off water?

1. Put a plant on cardboard.

2. Put a big glass jar over it.

3. Put clear jelly around the jar.

4. What do you think will happen?

5. Wait a few days. What is inside the jar?

6. Where did it come from?

Petroleum
Jelly

4.

FLOWERS AND SEEDS

Flowers for sale! Flowers for sale! People like flowers for their colors, shapes, and smells. **Flowers** are parts of plants. Not all plants have big, bright flowers. But all flowers have the same job. How do flowers help the plant?

A flower's job is to make **seeds**. This is the flower of a daisy. The seeds are made in the center of the flower. The seeds will fall to the ground. They will grow into new plants.

Each of these seeds will grow into a plant. Daisy seeds grow into daisy plants. Corn seeds can make only new corn plants. Seeds have different shapes, sizes, and colors.

Some flowers make a cover for their seeds. The cover is called a **fruit**. Some fruits have many seeds inside. Other fruits have only one seed.

Many animals like to eat the fruits of plants. All of these are fruits. They all have seeds. Can you name the fruits? What fruits do you eat?

Some flowers make fruits that are hard and dry. These fruits also have seeds inside. These seeds are good to eat. They are peanuts.

An acorn is a hard, dry fruit. Inside the acorn is a seed. The seed will grow into an oak tree. What animals like to eat the seeds of oak trees?

205

ACTIVITY

How are seeds different?

1. Collect many kinds of seeds.

2. Dry them on paper towels.

3. Glue them to a card.

4. Write the name of the plant each seed came from.

5. Draw a picture of the fruits that the seeds came from.

6. Put the seeds in groups. What is the name of each group?

PEOPLE AND SCIENCE

Many people like to garden. Gardeners buy seeds to start their gardens. Where do the seeds come from?

The seeds are grown on special farms. The seed farmer picks the fruits when they are very ripe. The fruit is mashed up. Then the seeds are taken out and dried. They are put in packets and sold.

Main Ideas

- The roots, stem, leaves, and flowers are parts of plants.

- Each part has its own job.

- Seeds and fruits are made by the flowers.

- We eat many parts of plants.

Science Words

Match each word to a part of the plant.

fruit	**leaf**	**stem**
root	**flower**	**seeds**

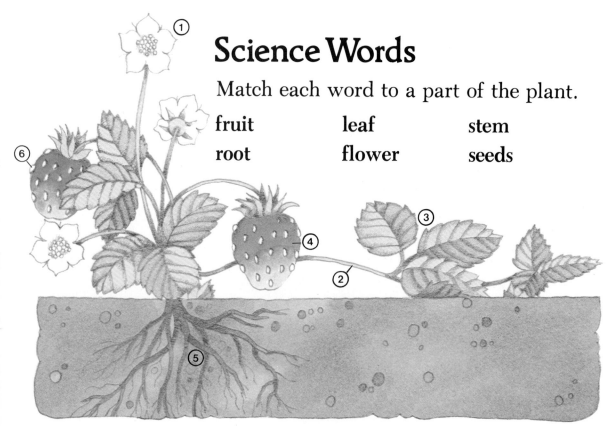

Questions

1. How does a plant get water?

2. Which tree here can make food?

3. What part of a plant makes seeds?

4. What do leaves use to make food?

5. Look at the picture. Which foods are leaves?

6. Which foods are roots?

7. Which foods are fruits?

CHAPTER 12

PLANTS ARE LIVING THINGS

1.

WHAT DO PLANTS NEED?

At this school the children work in a
greenhouse. A greenhouse is a place
where green plants grow. Green plants
are **living things**. The children learn
what green plants need to stay healthy.

All green plants need sunlight to live. The sunlight helps green plants make food.

In what part of the plant is food made?

Green plants need air to live. Some green plants live in water. They use tiny bits of air that are mixed in with the water. The air is used to make food.

Green plants need water to live and
to grow. They use water to make food.
What happens to plants when there is
no water? How do they look?

Most green plants need soil to live.
Soil holds water for the plants to use.
The soil has bits of rock in it. The plant
uses the bits of rock to make its food.

ACTIVITY

Do plants need sunlight and air?

1. Cover a leaf with black paper. The leaf will not get sunlight.

2. Cover another leaf with a plastic bag. This leaf will not get air.

3. Write down what you think will happen to the leaves.

4. Wait a few days.

5. Uncover the leaves. How do they look now?

6. Do plants need air and sunlight?

2.

WHERE PLANTS GROW

A dandelion grows on a lawn. The yellow flower changes. It turns white and fluffy. The wind blows the seeds. Tiny hairs help them stay up in the air.

Soon a seed lands. If it falls on soil, it will grow. On what other places could it land? Would it grow there?

This maple seed was carried by the
wind. It has a long wing. The wing
helped it to fly. The seed landed on soil.
There was some water in the soil. There
was air around it. The seed started to
grow into a tree.

Plants can grow indoors. The children give the plants soil, water, and light.

Some people grow plants around their homes. They plant grass seeds. They put whole plants in the soil.

What happens to soil when it rains? These students are trying to find out. Each tray of soil is like a hillside.

The roots of plants hold the soil. When it rains, the soil stays in place. Some hills have no plants. What happens when it rains? People grow plants on hills to hold the soil.

ACTIVITY

How do seeds grow?

1. Put a paper towel in a cup.

2. Put a little water in the cup.

3. Place bean seeds between the paper and the cup.

4. Keep the towel damp. Watch the seeds grow.

5. Draw a picture of them every day.

6. Which part grows first?

3.

IS IT A PLANT?

Around the trees are small, green
plants called **moss**. Moss grows in damp,
shady places. It grows in the forest. It
can grow in the cracks of sidewalks.

Moss plants grow close together. Did
you ever feel them? They feel like a soft
rug. Moss does not have flowers or
seeds. But it is a plant.

These plants are not green. They
cannot make their own food. They grow
on dead plants and animals. Others
grow on soil. They are called **fungi**.
Which kinds of fungi have you seen?

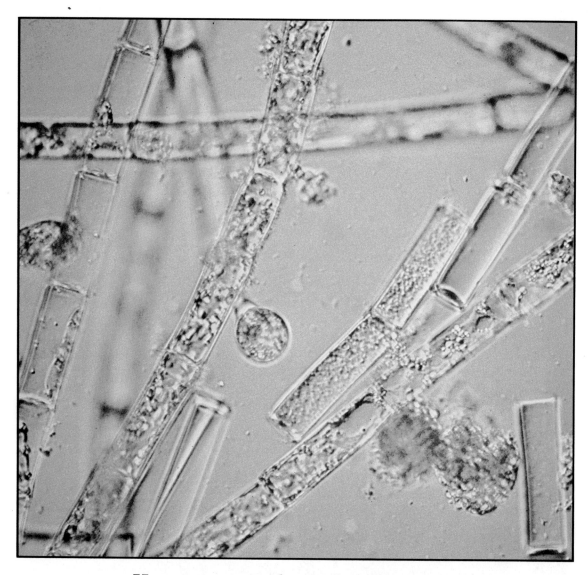

Here are some plants that live in a pond. They are very tiny. This picture was taken through a microscope. These plants are called **algae**. They do not have roots, stems, leaves, or flowers. These algae are green plants. They make their own food. How are algae different from other plants?

Some plants act like animals. This plant gets food in two ways. It is a green plant. It can make food in its leaves. But it can catch food, too. It is called a Venus' flytrap.

The leaves of a Venus' flytrap have tiny hairs. Insects touch the hairs. Then the leaves close tightly. The plant uses the insects for food.

ACTIVITY

Where do fungi grow?

1. Rub some dust on bread.

2. Put it in a plastic bag.

3. Add a slice of apple.

4. Put 10 drops of water in the bag.

5. Keep the bag in a warm place. Wait 3 days.

6. What do you see on the food?

7. Is it alive? How do you know?

Do you eat mushrooms? Mushrooms are plants. They do not need sunlight. They cannot make their own food.

A cave is a very good place to grow mushrooms. It is dark and damp. The cave is cool all year. The farmer grows the mushrooms in trays. The trays are filled with dead plant and animal materials. The mushrooms get food from them.

Main Ideas

- Plants are living things.

- Plants need air, sunlight, soil, and water.

- Plants grow in many places.

- There are many kinds of plants.

Science Words

Use these words to fill in the blanks.

living things   **soil**   **fungi**
moss   **algae**

1. The roots of plants hold on to _____.

2. _____, _____, _____ are plants.

3. All plants are _____ _____.

Questions

1. Which objects are plants?

2. Can a green plant live in a cave?
   How do you know?

3. The park changed. How did it
   happen?

Science Project

Plant a kitchen garden. Put some soil
in an egg carton. Plant some herb seeds
in the soil. Put your garden near a
sunny window. Keep the soil damp. Use
your plants for cooking.

SCIENCE WORD LIST

The science words are in ABC order. The number tells you on which page to find the word.

PHOTO CREDITS

TABLE OF CONTENTS:

p. *iv:* Chapters 1 – 3: Tom McHugh/Photo Researchers, Inc.; N.Y. Zoological Society; Frank J. Miller/Photo Researchers, Inc.

p. *v:* Chapters 4 – 6: Neil Leifer/Sports Illustrated; Scott J. Witte; Randy O'Rourke/The Stock Market

p. *vi:* Chapters 7 – 9: Derek Berwin/The Image Bank; Focus on Sports; Lee L. Waldman/1982/The Stock Market

p. *vii:* Chapters 10 – 12: Al Grotell; Tom and Michele Grimm/After Image; Niki Ekstrom

HRW Photos by Bruce Buck appear on pages 12, 68.

HRW Photos by Russell Dian appear on pages: 4, 8, 21, 26, 30, 31, 39, 47, 52, 61, 65, *bottom* 82, 84, *bottom left*, 85, 88, 92, 93, *bottom* 97, *bottom* 99, 100, 103 (N.Y.U. Theater); 104; 109, 113 (N.Y.U. Theater), 124, 128, 129, 133, *Inset* 141, 148, *right* 149 (Courtesy of the N.Y.C. Fire Dept.), 156, 160, 164, *top* 166, 176, 180, 192, 196, 197, 201, 206, 213, 214, 219, 224.

HRW Photo by Louis Fernandez appears inset on page 207.

HRW Photos by Henry Groskinsky appear on pages 77, 80, 101, 102.

HRW Photos by Richard Haynes appear on pages 110 and 111.

HRW Photo by William Hubbell appears on page 142 *top.*

HRW Photos by Richard Hutchings appear on pages: 58, 90, *top* 99, 125, *bottom* 126, *bottom* 149, 184.

HRW Photo by Ken Karp appears on page 131.

HRW Photos by Ken Lax appear on pages: *bottom* 41, 43, 44, *middle* 49, *bottom* 62, 64, 71, 78, 79, 81, 86, 89, 91, 106, 107, 108, 112, *bottom* 119, 122, 123, *top* 126, 127, 132, *left* 138, 139, 140, *bottom* 142, 143, 144, 146, 147, 159, *top* 175, *bottom* 183, 191, 195, 200, 203, 204, 211, *bottom* 212, 217, 218.

HRW Photo by Yoav Levy appears on page 153.

HRW Photo by David Lokey/Vail Associates appears on page 73 *left.*

HRW Photo by John Running appears on page 118 *bottom.*

HRW Photo by Scott Witte appears on page 13.

Chapters 1 – 6: p. viii — Tom McHugh/Photo Researchers, Inc.; p. 1 *top left* — Bob and Clara Calhoun/Bruce Coleman, Inc., *top right* — Laura Riley/Bruce Coleman, Inc., *bottom left* — Leonard Lee Rue III/Bruce Coleman, Inc., *bottom right* — L. Riley/Bruce Coleman, Inc.; p. 2 — Leonard Lee Rue III/Animals, Animals; p. 3 *top* — Ardea Photographers, *bottom* — Breck P. Kent/Imagery; p. 5 — Stephen J. Krasemann/Peter Arnold, Inc., p. 6 *top* — Unal Berggren/Ardea Photographers, *bottom* — Hans Reinhard/Bruce Coleman, Inc., p. 7 *top left* — Ronald G. Austing/Bruce Coleman, Inc., *top right* — Max Hunn/N.A.S./Photo Researchers, Inc., *bottom* — Alan Blank/Bruce Coleman, Inc.; p. 9 *left* — Dr. E. R. Degginger/Bruce Coleman, Inc., *right* — Weaver H. Muller/Peter Arnold, Inc.; p. 10 *top* — A. Grattell, *bottom* — Dr. Wm. J. Jahoda/N.A.S./Photo Researchers, Inc.; p. 11 *top* — Stan Leiser, The Sea Library, Photo Quest, *bottom* — Zig Leszczynski/Animals, Animals; p. 16 — N.Y. Zoological Society; p. 17 *left* — Gary Meszaros/Bruce Coleman, Inc., *right* — Karl Hentz/The Image Bank; p. 18 *top* — © 1973 Jen and Des Barlett/Bruce Coleman, Inc., *bottom* — Ardea Photographers; p. 19 *top and middle* — Breck Kent/Imagery, *bottom* — D. Hughes/Bruce Coleman, Inc.; p. 20 *top* — Breck Kent/Imagery, *bottom left* — Lynn M. Stone/Animals, Animals, *bottom right* — Ardea Photographers; p. 22 — D. Lyons/Bruce Coleman, Inc.; p. 23 *top* — Dr. E. R. Degginger/Bruce Coleman, Inc., *bottom* – -L. West/Bruce Coleman, Inc.; p. 24 *top and bottom* — Dr. E. R. Degginger; p. 25 *top and bottom* — Dr. E. R. Degginger; p. 27 — James H. Carmichael/Bruce Coleman, Inc.; p. 28 *left* — Lois and George Cox/Bruce Coleman, Inc., *right* — Dr. E. R. Degginger; p. 29 *top to bottom* — John Shaw/Bruce Coleman, Inc., L. West/Bruce Coleman, Inc., D. Overcash/Bruce Coleman, Inc., John Shaw/Bruce Coleman, Inc.; p. 32 — Hubertus Kanus/Shostal Associates; p. 35 *left* — Phil Degginger/Bruce Coleman, Inc., *right* — Mark Hanaur; p. 38 — Frank J. Miller/Photo Researchers, Inc.; p. 40 — John Zimmerman/Sports Illustrated; p. 41 *top* — Daniel Brody/Editorial Photocolor Archives; p. 42 *top* — Norman Tomalin/Bruce Coleman, Inc., *bottom* — Grant Heilman; p. 45 — Ed Gallucci/The Stock Market; p. 48 — Frank Whitney/The Image Bank; p. 49 *top* and *bottom* — Grant Heilman; p. 50 *top* — Grant Heilman, *bottom* — W. Bayer/Bruce Coleman, Inc.; p. 53 — Roger Tully/Black Star, *inset* — Lee Balterman; p. 56 — Neil Leifer/Sports Illustrated; p. 57 — Bullaty Romeo; p. 59 — Dennis Brack/Black Star; p. 62 *top* — Larry Dale Gordon/The Image Bank; p. 63 — Jonathan Wright/Bruce Coleman, Inc.; p. 66 *top* — Grant Heilman, *bottom* Mort Beebe/The Image Bank; p. 69 — Richard W. Brown/F.P.G./Alpha; p. 70 — Peter Arnold; p. 72 — NOAA; cumulonimbus: Bruce Coleman, Inc.; p. 73 *right* — Clyde H. Smith/Peter Arnold, Inc.; p. 76 — Scott J. Witte; p. 81 *right* — Webb Photo; p. 82 *top* — Four By Five, Inc.; p. 83 — Jim Pickerell/Black Star; p. 85 *top left* — James P. Rowan/Gartman Agency; p. 85 *right* — Leif Erickson/The

ART CREDITS